AIChE Equipment Testing Procedure

PARTICLE SIZE CLASSIFIERS

2nd Edition

Prepared by the Equipment Testing Procedures Committee
of the American Institute of Chemical Engineers

ISBN 0-8169-0594-0

Procedure E-29

American Institute of Chemical Engineers Equipment Testing Procedures Committee

Chair:

S. D. Fegan

ABS Lawrence Pump & Engine Co.

Vice Chair:

J. F. Hasbrouck

Union Camp Corp.

Particle Size Classifier Procedure Revision Subcommittee

Chair:

R. H. Snow

IIT Research Institute

Members:

P. T. Luckie
Pennsylvania State University

T. Allen
E. I. duPont de Nemours and Co.

Contributing Reviewers

T. Blackwood
Monsanto Chemical Co.

R. Karuhn
Particle Data Laboratories Ltd.

I. Klumpar
Badger Engineers

K. Leschonski
Technische Universitat Clausthal

B. Scarlett
Technische Hogeschool Delft

J. S. Summers
U. S. Bureau of Mines

B. Wood
Eastman Kodak Co.

Officially approved for publication by AIChE Council on April 29, 1993.

Other available
Equipment Testing Procedures
Include:

Packed Distillation Columns

Fired Heaters

Spray Dryers

Mixing Equipment (Impeller Type)

Tray Distillation Columns

Direct-Heat Rotary Dryers

Centrifugal Pumps

Centrifuges

How to Use This Procedure

This publication provides a procedure for testing and evaluating particle classifiers. The results are expressed as a classifier size selectivity function. This function, because it tends to eliminate the effect of the feed size distribution, primarily is a property of the classifier and its operating conditions. Selectivity then can be combined with the feed size distribution to calculate other measures of performance, such as yield and recovery.

The procedure includes methodologies for sampling and measuring particle streams, and summarizes methods of particle size analysis. It lists operating variables to be considered and measured. Although the procedure is intended specifically for particle classification equipment, as distinguished from particle collection devices, many of the items discussed also are relevant to collection equipment.

Because it is intended to apply to any type of particle classifier, this procedure only should be used by the engineer for guidance for preparing a specific set of instructions, following the principles set forth herein, for the particular classifier to be tested.

Note that the American Institute of Chemical Engineers (AIChE) and members of its Equipment Testing Procedures Committee make no representation, warranties, or guarantees, expressed or implied, as to the application or fitness of the testing procedure suggested herein for any specific purpose or use. Company affiliations are shown for information only, and do not imply approval of the procedure by the companies listed.

Contents

100.0 **Purpose and Scope**

101.0 *Purpose*

The primary *purpose* of this procedure is to provide methodology for conducting and interpreting performance evaluation tests on particle classification equipment. Emphasis herein is directed toward the equipment user in assessing performance relative to that person's own application. See also Sect. 301.0 — Objective of Testing Classifiers.

Another purpose is to adopt terminology and nomenclature consistent with the two major classifier application industries — namely, the chemical process and mining industries.

102.0 *Scope*

102.1 Included in the scope are the following general types of classifier equipment:

- those using gas flow, as in air separators and cyclones;
- those using liquid flow, as in hydrocyclones and screws; and
- those using solid flow, as in screens.

102.2 It is intended that the scope of this procedure be limited to classification by size or shape, *not* including separation by *type of material* or *composition*. For example, some classifiers, including the gas- and liquid-flow types, separate based on interaction of *inertial* and *drag forces*, and inertia is affected by density. Thus materials of different density in a mixture such as a crushed ore would separate at different sizes in such classifiers. These classifiers can make only a rough separation of such mixtures by either size or composition. They would separate purely by size only a material of uniform density, and by composition only a material of uniform size. Screens, on the other hand, separate primarily by a gauging action, relatively independent of density. See also Sect. 202.3.

102.3 This procedure is intended to be applied to particle size *classification* equipment as distinguished from particle *collection* equipment. The purpose of particle size classification is to separate particles by size; a collection system is designed to collect or recover as much particulate matter as possible from a gas or liquid stream irrespective of size. Many of the items discussed (especially sampling, particle size analysis and performance criteria), however, are also applicable to collection systems.

102.4 The major technical areas addressed herein are particle sampling, particle size measurement, and evaluation methodology.

102.5 This procedure will be useful to the engineer who intends to test and evaluate the performance of classifiers. Moreover, in providing a standard methodology and terminology, it should encourage the uniform and routine publication of usable performance data by equipment manufacturers, to aid equipment selection.

200.0 **Definitions and Descriptions of Terms**

201.0 *Classification*

201.1 *Classification* is the process whereby a collection of particles is separated into two or more portions differing in some physical or chemical property.

201.2 *Size classification* is a classification process in which the collection of particles is split into fractions comprising particles of different size ranges. For example, in a single-stage classification, particles below a stated size are separated from the feed stream to produce a fine stream containing most of the fine particles. The amount separated usually represents a major portion of the feed stream.

201.3 *Dedusting*, as contrasted with size classification, is a term used to describe the process of removing "dust producing" fine particles below a stated size from a feed stream, in order to produce a coarse stream essentially free of particles smaller than the stated size. Usually the amount to be removed represents a minor portion of the feed material.

201.4 *Fraction* is the term applied to each portion of a collection of particles resulting from a classification process. A fraction can be measured in terms such as mass, volume, number or surface area size distribution. Although a size classification process can yield a large number of fractions, such a process can be considered as a number of separations into a coarse and a fine fraction, each at a different separation size. In that case, however, it must be remembered that the feed material to each successive separation will be different in composition from that to the previous separation. In the subsequent discussion, it will be assumed that a single classification of a given feed material will be taking place to yield a single coarse and a single fine fraction.

202.0 *Particle Size*

202.1 *Particle size* is intended to describe a particle in terms of a characteristic linear dimension, such as diameter for spherical particles. It should be noted that in the U.S.A. it has been customary (although not universally so) to express size in terms of a diameter. Failure to specify radius or diameter has been a source of confusion.

202.2 With spherical particles, regularly shaped particles, and complex but similarly shaped particles, it is theoretically possible to make a separation on the basis of some linear dimension. Irregular particles, however, are of non-similar *shape*, and it is not possible to define a demarcation in terms of any specific linear dimension alone. Instead, size must be defined in terms of some other size-related property such as area or volume, or in terms of a statistically defined dimension.

202.3 In some cases, it has been customary (and theoretically rigorous) to specify the size of the particle as the size of an *equivalent sphere.* The significance of "equivalence" must then be remembered because the "equivalent size" can be different depending on the size-related property employed. The obvious property is mass, but there are other possibilities. Often a size separation is made on the basis of a process that is predominantly determined by size but also is influenced by other properties of the particle. For example, a separation can be made on the basis of settling velocity in a fluid, which is dependent on the particle size, shape, and density. In those cases, it has also been customary to specify size by the size of a sphere having the equivalent overall behavior in that kind of process. Such an equivalent size is a rigorously correct representation of size if the other properties of the particle are constant. If the other properties also vary among the particles of different size, however, this equivalent size is no longer representative of the size effect alone, but of the overall effect of all the relevant properties in terms of the specific process employed. Thus, in the general case, an equivalent size is a representation of size in terms of some specific phenomenon. In addition, the composition may vary with size and, therefore, the density also may vary with size. It is only when all particulate properties influencing the phenomenon, other than size, are constant,that an equivalent size can be taken as a direct measure of particle size alone.

203.0 *Particle Size Distribution*

203.1 *Particle size distribution* is a representation of the relative amounts of material as a function of size. There can be various distributions depending on the property used as a basis for designating "amount." The common bases for designating "amount" are: (1) number of particles; (2) total particulate area; and (3) mass or volume of particles. Actually, any number of bases could be used, including chemical or optical properties.

203.2 Size distributions may be given on either (1) a frequency basis, or (2) a cumulative basis. A *frequency distribution* reports the fractional amount (on some specific basis) of the total particles that lie in an incremental size range as a function of size, where the incremental size range is chosen on some systematic basis. There can be an infinite number of ways of presenting a frequency distribution depending on the system for choosing incremental size. A *cumulative distribution* gives the total fractional amount (on some specific basis) of particles that exist above or below a specific size. The cumulative distribution is essentially the integral of the corresponding frequency distribution. For a given basis of measurement and size discriminatory property, however, there is theoretically only one cumulative distribution regardless of the manner of depicting the frequency distribution.

204.0 *Particle Size Analysis*

204.1 *Particle size analysis* is the technique used to obtain a particle size distribution. There are two broad categories of size analyses: (1) those in which individual particles are measured and counted; and (2) those in which the material is separated into size fractions, the quantity of each then being measured.

204.2 Every size analysis involves two basic considerations: (1) the mechanism used to *discriminate* between particles of different size in either counting particles or in preparing fractions; and (2) the basis used for measuring the *magnitude* of the fractions obtained. Those categories of size analysis in which individual particles are counted are all effectively on a count basis, although the discriminating mechanism can be different for different methods of counting. In any case, the discriminating mechanism determines the significance of the equivalent size in which the distribution is reported. A full discussion of discriminatory mechanism, quantity measurement bases, representation of size, and methods for converting from one basis to another are given in Ref. 803.6.

205.0 *Classification Criteria*

205.1 *Classification criteria* are those items used to specify the conditions at which a separation is made and to measure the effectiveness of the separation.

205.2 *Feed* is the total amount of particulate material fed to a classifier to be separated into one or more fractions according to some specified characteristic of the particulates.

205.3 *Product* is the fraction (or fractions) that supposedly contain the material having the desired specified characteristics.

205.4 *Yield* is the total amount of material in the product. All of the yield does not necessarily represent material of the desired characteristics. It may be expressed as a fraction or percentage of the feed. (See Sect. 602.1.)

205.5 *Recovery* is the amount of material of the desired characteristics that is present in the yield. It is usually expressed as a fraction or percentage of the total amount of material of the desired characteristics present in the feed. For the perfect or ideal separation, the amount of yield and amount of recovery are equal; however, for such an ideal separation when expressed as a fraction or percentage of the amount of material in the feed, the recovery is 100% of the material of the desired characteristic in the feed, while the yield will depend on the amount of desired material originally present in the feed. (See Sect. 602.1.)

205.6 *Selectivity function* is the term applied to the measure of classification performance. In a size classification process, each particle size in the feed distribution has a certain probability of entering the coarse fraction. For example, smaller particles have a lower probability of entering the coarse fraction than larger particles. The curve representing this process is called a size selectivity function*, and expresses the probability of finding a particle of a stated size in the coarse fraction. (See Sect. 601.2.)

205.7 *Selectivity* of a classifier is generally a *function* of the *feed size distribution,* as well as other operating conditions. Yet its dependence on feed size distribution is an indirect effect, and this dependence may be a weak one. In contrast, other measures of performance described in Sect. 603. have the feed size distribution as a factor.

205.8 In order to obtain the *intrinsic* performance capability of a classifier, which we may term *"intrinsic" size selectivity*, it is necessary to operate with a completely dispersed collection of particles at low capacity so that particle interactions are avoided and the classifier can act on each particle as if others were

* Other terms that appear in the technical literature having identical meaning to size selectivity include grade efficiency, fractional efficiency, and particle size efficiency.

not present. Because in practice we have no assurance of complete dispersion, size selectivity should be regarded as a way of presenting classifier performance in a given application which incorporates the effects of material, capacity, and intrinsic classifier properties, and may be presented as a function of these properties.

205.9 *Cut size* is the name applied to the equivalent size at which a classification is being made. For a perfect separation, this size is determined unequivocally. For a less than perfect separation, however, a further definition is needed, and this leads to a large number of possibilities. A common definition of cut size, referred to as equiprobable cut size, is that feed particle size having equal probability of entering either coarse or fine fraction. (See Sect. 601.5.1.)

205.10 *Sharpness* is the term applied to the effectiveness of a given classification. There are a large variety of ways in which sharpness can be expressed. For a less than perfect separation, the method of specifying sharpness and the method of defining cut size may be interrelated. Sharpness is related to the shape of the size selectivity curve. (See Sect. 601.5.)

206.0 *Dispersion*

206.1 *Dispersion* is the act of distributing the particles in a medium so that the particles exist as separate entities not adhering to each other. *Degree of dispersion* is the extent to which the dispersion process is effective. Particles which are not dispersed are said to be flocculated, agglomerated, or aggregated. These terms are defined in Ref. 803.9, p. 8–57, and are summarized below.

206.2 An *agglomerate* is any collection of particles bound together by physical or chemical forces. These forces can include intermolecular and electrostatic forces in very fine agglomerates, capillary forces when liquids are present, and bridges of dried or sintered solids.

206.3 A *floc* is a weakly bound assemblage that may be temporary if the binding forces are not sufficient to resist inertial forces.

206.4 The term *aggregate* should be reserved for a collection of particles of differing composition.

300.0 **Test Planning**

301.0 *Objective of Testing Classifiers.*

The objective of a test program is to establish the effectiveness of a classifier. The following lists the possible objectives of such a test program:

301.1 Tests can be performed on existing installed classifiers to check their *operating performance*.

301.2 Size selectivity data from tests can be published to aid users to *select equipment*, and to further equipment development.

301.3 Tests can be performed on laboratory classifiers to obtain data necessary for *scaleup*, and to estimate capital and operating *costs* for commercial sized plants.

301.4 Sample products can be obtained with a selected range of particle size distribution, and these can be used to *evaluate product quality* and performance in some application.

302.0 *Testing Instructions*

302.1 General instructions are dealt with in this section. For specific instructions, see Sect. 500.0. Because this procedure is intended to apply to any type of particle size classifier, its instructions cannot be specific to any one type. Therefore, the engineer using this procedure must prepare specific instructions, following the principles set forth herein. A summary of factors these instructions should consider is given below.

302.2 *General Factors to be Considered*

302.2.1 *Objectives* of the test for a specific application.

302.2.2 *Variables* of classifier operation to be measured, those to be controlled, and those to be varied.

302.2.3 Method of *presenting results* of tests.

302.3 *Specific Factors to be Considered*

302.3.1 Method of *feeding* a uniform particulate stream to the classifier.

302.3.2 *Duration* of each test period.

302.3.3 Method of *sampling* feed and fraction streams.

302.3.4 Method of measuring both particle *flow rate* and fluid flow rate of feed and fraction streams.

302.3.5 Method of *subdividing* samples for analysis.

302.3.6 Method of *particle size analysis* to be used.

302.3.7 *Statistical experiment design* principles as discussed for example in Ref. 803.13 and summarized in Sect. 508.

302.3.8 *Safety* and *environmental* precautions.

303.0 *Factors and Conditions to be Recorded*

303.1 Because a classifier will make products over a wide range of operating conditions, a program to completely test its performance should measure the effect of operating variables on product particle size and yield over the full range of operating variables available to the classifier under test. A summary of the type of information that will be required in the general case is presented below. All this information will not necessarily be applicable for all types of classifiers.

303.2 *Equipment Description*

303.2.1 General dimensional details

303.2.2 Mechanical setting of variable items (such as vanes, fingers, baffles, etc.)

303.2.3 Auxiliary equipment (feeders, conveyors, drives, etc.)

303.2.4 Materials of construction

303.3 *Operating Conditions*

303.3.1 Mechanical

- Speed of rotation
- Power consumptions of drives
- Vibration frequency and amplitude

303.3.2 Fluid

- Flow rates, including primary stream, secondary stream, and recycle streams.
- Condition (temperature, pressure, humidity)
- Pressure drops and power consumption
- Additives used (dispersion agents)

303.3.3 Particulate

- Feed rates, magnitude and degree of uniformity
- Dispersion methods
- Material balance (coarse fraction, fine fraction, recycle, losses)
- Condition (temperature, moisture content)
- Contamination of product stream
- Physical properties of feed and fractions (size distributions, specific gravity, etc.)

303.4 *Operating Problems*

303.4.1 Health and safety related (toxicity, explosive, ignition, static charge)

303.4.2 Product quality (grinding, contamination)

303.4.3 Maintenance

- Mechanical
- Corrosion
- Wear
- Plugging

303.5 *Analytical Procedures*

303.5.1 Sampling

303.5.2 Rate measurement

303.5.3 Particle size measurement

304.0 *Safety and Environmental Precautions*

304.1 Before embarking on a test program, personal safety requirements should be considered. Feed characteristics to be considered are *toxicity*, *explosiveness*, *minimum ignition energy*, and *static charge generation*. After analyzing safety hazards, protective measures such as respiratory or other personnel protective equipment, equipment grounding and bonding, protection from tramp metal, inert gas blanketing, or exhaust ventilation can be provided.

304.2 For information on prevention of dust explosions, consult, for example, Ref. 803.9, Sect. 8, pp. 11–12. One should be aware that *aerosols of any combustible material*, such as starch, sugar, or powdered metals, can be *explosive*.

304.3 The test procedure should conform to the latest requirements of all applicable *safety standards*. These include, but are not limited to, plant, industry, local, state, and federal regulations. The plant management (or the equivalent if the test is not run in a plant) and the classifier manufacturer should be requested to furnish —in writing — specific details that should become a permanent part of the test record. It is recommended, further, that all testing be conducted by, or under the supervision of personnel fully experienced in plant and equipment operating procedures.

304.4 The test procedure must conform to the latest requirements of all applicable *environmental standards* which include plant, industry, local, state, and federal regulations. Environmental conditions that apply to the equipment in normal operation should also apply during testing.

400.0 **Instruments and Methods of Measurement**

401.0 *Introduction*

401.1 The two most important measurement methods used in evaluating the performance of classifier equipment are *particle sampling* and size analysis. Particle sampling is an important aspect because particle segregation can lead to biased information. *Particle segregation* means that particles with different properties are located in different positions. For example, when particulate material which is prone to segregate is poured into a pile, the larger particles tend to roll farther than the finer ones. Therefore, the use of a scoop or a thief probe in this pile will give different results, depending on where the sample is taken. Even riffling methods, such as coning and quartering or the multi-chute riffler, that subdivide the sample into supposedly equivalent portions can only partially overcome the variations in composition caused by segregation.

401.2 A *size analysis* of classified materials may be desired for either of two purposes: (1) to establish the intrinsic classification effectiveness of a given classifier, *i.e.*, its design performance; and (2) to establish the effectiveness of classification for a *specific application*, which may be influenced by specific feed characteristics.

> 401.2.1 For evaluating *intrinsic classifier performance*, it is desirable to use a size analysis method or combination of methods that are accurate over the whole range of feed sizes expected. For example, a particle counting method may meet this requirement if a statistically sufficient number of particles is counted in each size range of interest, especially at the coarse end.
>
> 401.2.2 For evaluating classification effectiveness for a *specific application*, it is desirable to use a size analysis method sufficiently accurate for the use intended. If the size analysis method uses a separation mechanism or measures a physical property that indirectly measures size, then this should be similar to that involved in the process for which the product is to be used (*e.g.*, surface area for a process involving surface chemical reaction), regardless of the principle of separation employed by the classifier. Thus, any significant mechanistic differences between the classifier and the

application will show up as a poorer apparent classification than would otherwise be possible. This, however, is a realistic measure of the results being achieved for the application desired.

401.3 For reasons of economics, practicality, convenience, or specification, it is often not feasible to use the most desirable or realistic size analysis technique. If the most desirable size analysis technique for the purpose cannot be used, an *intrinsic discrepancy* in apparent effectiveness of classification is introduced by the size analysis itself and must be recognized as such. In the present state-of-the-art, the specific mechanistic role of particle size in most applications is not understood adequately to permit quantitative assessment of the types of size analysis discrepancies discussed above except for obvious extreme cases.

402.0 *Particulate Sampling*

402.1 The most effective method of obtaining a *representative sample* depends on passing the entire stream through a device which will cut portions from the stream at successive time intervals. Using this method, a segregation occurring across the stream has no effect on the sampling. Further, any variation in composition of the stream with time will be averaged and represented in the sample in an unbiased way, provided that the frequency of cuts is statistically sufficient. The following methods of sampling are based on these principles. (See also Refs. 803.1, 803.2, 803.3, and 803.5.)

402.2 *Sampling Particles from Conveyors or Feeders*

402.2.1 Frequently the feed stream to a classifier is fed from a hopper through a conveyor and is already in the form of a stream of solid particles. It is only necessary to insert an appropriate *sample cutter* at a convenient point where the whole stream can be cut — for example, the discharge of the feeder into an intake chute, or a transfer point between conveyors. Feed systems should be designed to provide a sampling position.

402.2.2 Two types of commercially available sample cutters are shown in Figure 4.1 (Ref. 803.9, Sect. 21, pp. 36–39). In Figure 4.1a, the cutter is a box with a discharge chute. The top of the box has a rectangular opening. On signal from a timer, a motor draws the cutter across the sample stream. The cutter speed must be uniform to assure that all portions of the stream cross-section are sampled at the same rate. The cutter opening must be large compared to the size of particles to be sampled, so that edge effects do not bias the sampling of the largest particles.

402.2.3 Figures 4.1b and 4.1c show *rotary samplers* or *rifflers* operating on a similar principle. These devices contain a cutter with rectangular cross-section, mounted on a box that rotates around an axis. The feed stream passes through a vertical chute and is intercepted periodically as the cutter rotates into the feed stream at a uniform speed. These rifflers are designed to extract a sample amounting to either 5 or 10% of the feed stream. Further reduction of the sample is done by passing it through subsequent smaller stages of the same type of riffler.

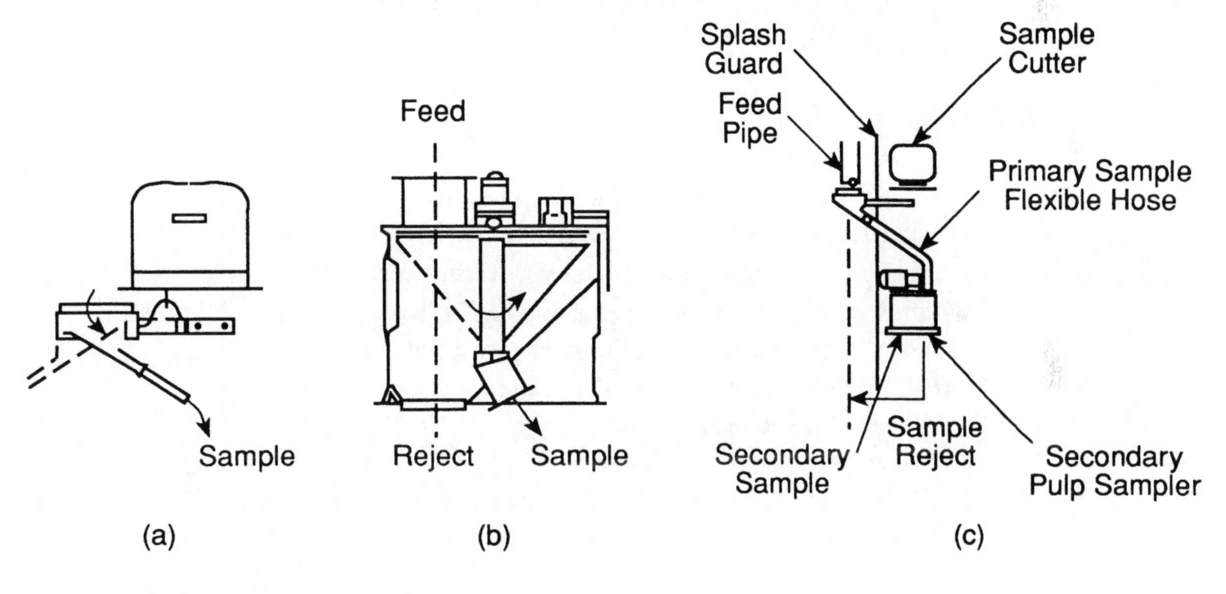

(a) Linear Cutter
(b) In-line Rotary Sampler
(c) Primary and Secondary Sample Reduction

(b and c courtesy of Denver Equipment Co.)

Figure 4.1. Types of mechanical samplers.

402.3 *Sampling Bulk Powder*

402.3.1 In some cases, the particulate material is in a pile or a container. Since particles usually *segregate* when poured onto a pile, it is difficult to provide a uniform, nonvarying feed stream from a pile. Although uniformity can be improved by recovering the material by cutting through the pile, as is done in coning and quartering as described below, there still will be some nonuniformity. Therefore, it is important to adequately sample and measure the stream as it varies over time.

402.3.2 *Feed sampling* can best be done by passing the feed materials through a conveyor or feeder and sampling by use of the types of cutter samplers described above. Preferably, this will be done while feeding the material to the classifier during the test, because any variation in composition while feeding during the test will be reflected in both inlet and outlet samples and thus compensated for.

402.4 *Sample Reduction*

402.4.1 Depending on the scale of operation of the system being tested, the samples may range in size from a few grams to many kilograms. The larger samples must be further reduced to provide *representative samples* of the quantity needed for size analysis. This can best be done using the principles discussed above. The rotary riffler shown in Figure 4.1b can reduce a large sample. It is only necessary to feed the material to the riffler at a reasonably constant rate. This can be done by placing it in the hopper of a table or belt feeder. If the material is sufficiently free flowing, a feed funnel can be used.

402.4.2 Once the sample is reduced to a few kg, it can be further subdivided on a *laboratory rotary riffler*. Figure 4.2 shows the operating principle of a spinning riffler which divides the sample into as many portions as there are receiving containers. Further subdivision is achieved by repeat riffling. If there are ten receiving containers, three passes will reduce the sample size to 1/1000 of the initial sample.

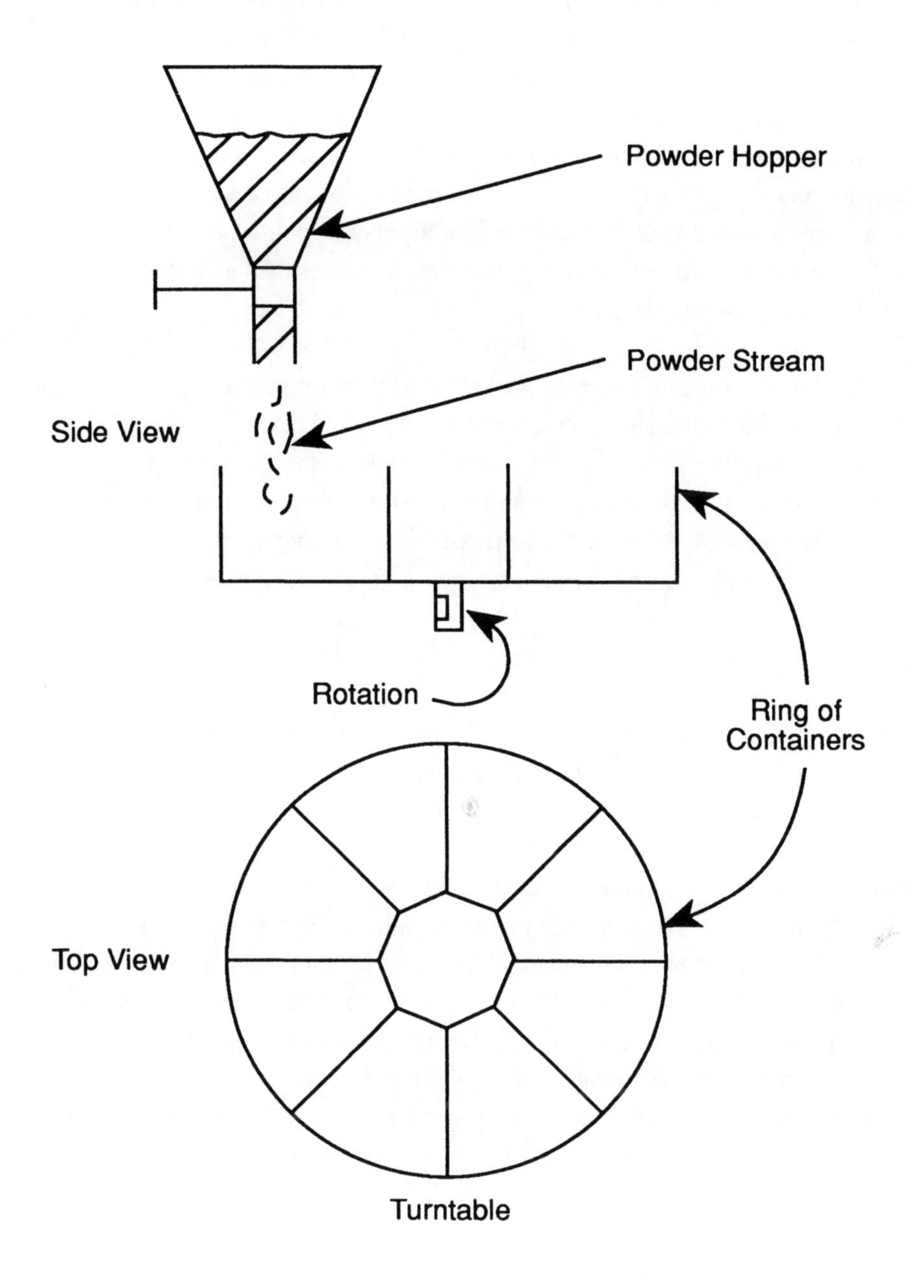

Figure 4.2. Spinning riffler sampler.

402.4.3 The *multi-chute sampling riffler* commonly seen in particle laboratories is less effective than the spinning riffler. It only achieves a 1:2 reduction on each pass, and it cuts the sample by location rather than by time. Segregation occurs during pouring the particles onto the riffler. If the sampler has a sufficient number of chutes, it can give a representative sample anyway; the commercially available units do not have a sufficient number of chutes, however. The use of a *spinning riffler* is recommended rather than a chute sampler. (See Ref. 803.5.)

402.4.4 *Coning and quartering* is reported to be an accurate method of obtaining a representative sample, if it is carefully done, but it is too laborious for ordinary use. In this method the material is piled on a firm floor, and cut through to divide the pile into pie-shaped quarters. Because the larger particles roll to the edge of the pile, the sample will be biased if the cutting is not done carefully.

402.5 *Sampling Particles from Slurries*

A number of industrial processes, particularly in the ore processing and cement industries, require classification of particles in slurry form. Examples of such classifiers are hydrocyclones, wet screens, and settlers. The method of sampling a slurry stream is the same as that used for sampling a powder stream — using rifflers that cut across the stream repeatedly. By these means, one can sample streams of high volume, up to approximately 1 ton/min (20 kg/sec) or more. Sometimes the slurry stream is accessible at a transfer point, such as where it flows out of a ball mill. Unfortunately, it is frequently the case that the plant was not designed with sampling in mind — *e.g.*, the slurry stream is pumped through large-diameter pipe with no transfer points where the whole stream can be sampled. It may be necessary to cut out a section of pipe and insert a riffler of the type shown in Figure 4.1c. The feed chute of the riffler then becomes a section of the slurry pipe.

402.6 *Sampling Particles from a Gas Stream*

402.6.1 In many industrial applications, the particles to be classified are suspended in an air or gas stream. Where the feed and fractions are in bulk form, the methods for bulk powder sampling given in Sect. 402.3 can be used. More often, the feed and at least one fraction stream are associated with some other operation, such as a closed circuit grinding mill. In this case, it is necessary to extract representative samples from the gas streams.

402.6.2 If the scale of operation is small (less than 700 ft^3/min or ⅓ m^3/sec), a small bag filter could be used to avoid the sampling problem. Most industrial operations are larger than this, so it is not practical to divert the whole gas stream into the sampler. Then the only practical method is to sample by means of a *probe.* In this case, best efforts must be made to minimize nonrepresentative sampling due to particle segregation within the gas stream.

402.6.3 *Particle terminal settling velocity* and *concentration* are variables that determine the extent of segregation that may occur. The terminal settling velocity of particles is a function of their size and density. If the terminal settling velocity is small compared to the turbulent component of velocity in the duct, then segregation due to settling will be minimized.

402.6.4 If the *particle concentration* in the gas stream is high, less segregation will occur because the more frequent interparticle contacts reduce the tendency of particles of different mass to flow past each other. The limiting concentration for this effect, however, has not been determined.

402.6.5 For coarser particles, segregation due to *settling* in the duct may be a problem. Ducts should be designed so that velocities are well above conveying velocities (typically over 50 ft/sec or 15 m/sec, but this depends on the size and density of the particles) in order to minimize gravitational segregation.

402.6.6 Size segregation can also occur at the entrance to a sampling probe if the flow velocity in the probe is not *isokinetic* with the flow in the duct due to inertial differences between small and large particles. Isokinetic sampling is important if particles coarser than 5 μm diameter are present in the stream and especially if the particle concentration is dilute. Isokinetic sampling is achieved by adjusting the flow velocity in the probe entrance equal to that in the duct. This is a difficult task, and it is preferable to use one of the other methods of sampling discussed above. Sampling segregation can be minimized by using large sampling probes with correspondingly high sampling rates.

403.0 *Size Analysis Methods*

403.1 *General*

403.1.1 In general, any method of size analysis can be employed to determine *selectivity* in order to assess classifier performance. Choice is often dictated by availability, capital costs, and operating costs. For classification processes or applications, the *mass distribution* is usually of prime interest; hence methods that give the mass distribution directly are preferred over those that give *number distributions*, because of the inherent errors in converting a number distribution to a mass distribution. For example, the omission of a single 10 μm particle in a distribution having a ten to one spread in size is equivalent to omitting one thousand 1 μm particles. Thus, the narrower the size distribution, the less the potential error.

403.1.2 The following describes some of the methods that have been used, with a commentary on their applicability and limitations. For a more detailed discussion of size analysis methods, see Ref. 803.18. *Note that trade names and manufacturers are cited to aid the user in locating available equipment. Additional vendors may offer similar, or otherwise suitable devices. Mention of a specific company or technique does not imply endorsement by AIChE or the Equipment Testing Procedures Committee.*

403.2 *Sieving*

403.2.1 The most common method of size analysis involving classifying powders into size ranges is by *sieving*. Heavy duty sieves are often made of perforated plate giving rise to circular holes; various other aperture shapes such as slots for sieving asbestos fibers are also available. Fine sieves are usually woven with phosphor bronze wire, medium with brass, and coarse with stainless steel. The sieve size is often specified by mesh number, which refers to the number of openings per linear inch. Thus the 200-mesh sieve contains 200 wires per linear inch of 50 μm diameter wire, leaving nominal 76 μm square apertures between the wires, and an open area of 35%. Each mesh number refers to a specific aperture, because the wire size corresponding to each sieve mesh is fixed by the sieve standards, decreasing with increasing mesh number. In general, the percentage open area decreases with increasing mesh number, thus prolonging the sieving rate for finer powders.

403.2.2 The standard sieve scale in the U.S.A. is the *U.S. Sieve Scale*, although the Tyler series is still used. These standard mesh numbers are limited to sieves having apertures commencing at 37 μm and then increasing in a fourth root of two progression of sizes. Electroformed sieves down to 5 μm are also available. Because of particle adhesion problems, mechanical sieving may not be possible for very fine powders. In such cases, the use of dry dispersing agents may facilitate sieving. Alternatively, *sonic sifting* (ATM Sonic Sifter), *air jet sieving* (Alpine Air Jet Sieve), or *wet sieving* can be used. A variety of wet sieving procedures and apparatus have been described.

403.2.3 It is emphasized that for reproducible data a *specified sieving procedure* such as an ASTM or International Standards Organization (ISO) standard method should be followed; it is essential for comparable data from different sets of sieves to calibrate the sieves. Results for mechanical sieving depend on mode of vibration, duration of sieving, and sieve loading. Sieves should be regularly checked for wear and damaged sieves discarded.

403.3 *Air Sedimentation or Elutriation*

The rate at which particles settle in air under the influence of either a gravity or centrifugal field is a function of particle size, shape, and density. This appears superficially to be an excellent way to size powders in order to assess classifier performance because it mirrors the behavior of the particles in the classifier. In the previous edition of this report, the *Roller Analyzer*, the *Micromerograph* and the *Bahco* are described. These instruments now are rarely used because of difficulties with dispersion that lead to poor reproducibility, protracted analysis times, high operating costs due to operator involvement, and, most importantly, the hazards due to dust escaping into the working environment. Currently, only the Bahco is commercially available.

403.4 *Liquid Sedimentation or Elutriation*

This method can be used for particle size analysis in a manner comparable to air sedimentation or elutriation. Liquid elutriation is rarely used, however, because of problems associated with secondary flow patterns.

403.4.1 Other wet sedimentation systems are used and require that the powder be dispersible and relatively insoluble in the sedimentation liquid. *Dispersion* is the most critical parameter in obtaining accurate and reproducible data with liquid sedimentation techniques and may be achieved by a combination of chemical and physical means. *Dispersing agents* may be cationic, nonionic, or anionic; these coat the particles, and may operate by *steric interference* or so that the particles mutually *repel* each other. Adjustment of the *zeta potential* to 30 mV or more also may aid dispersion. Mechanical agitation usually is necessary to incorporate the powder into the liquid by breaking up agglomerates; this may be achieved by spatulation, stirring, blending with a Waring-style blender, or ultrasonic means using a bath or probe. *Magnetic powders* or ores (such as magnetite) need to be magnetically deflocculated (depolarized) by alternating current fields before proceeding with sample preparation. It is essential that the *quality of the dispersion* be determined, usually by microscopic examination, before an analysis is carried out.

403.4.2 The applicable size range for gravity sedimentation in water is about 1 μm to 100 μm, but this can be extended at the high end by the use of more viscous liquids. Reasonable data can be generated at smaller sizes provided the size range is narrow.

403.4.3 The *Andreasen pipet* method is a sedimentation technique in which a quiescent suspension of particles, at a volume concentration of less than 1% (some authorities, such as Ref.

803.19, specify 0.2%) is sampled at a fixed level in a settling column by means of a pipet. The size distribution is obtained from data of the powder concentration as a function of sedimentation time and height of fall. Powder concentration normally is determined by drying and weighing the extracted samples after correcting for the weight of the dispersing agent. (For hygroscopic agents, it is necessary to dry in a desiccating chamber.) The procedure is versatile and reproducible but laborious with a lower limit of around 1 μm.

403.4.4 The *Ladal pipet centrifuge* is a centrifugal version of the Andreasen pipet that extends the lower limit into the submicron range (Ref. 803.20).

403.4.5 The *PAAR Lumosed* is a gravitational sedimentometer having three light beams at depths of 1.5 mm, 15 mm, and 150 mm. The signals from the three detectors are blended, with the fine end of the size spectrum determined from the first detector, the intermediate range from the second, and the coarse end from the detector at the greatest depth. Particle size is determined from the Stokes equation, and the concentration undersize from the light attenuation. Due to the complex interaction between the particles and the light beam, empirical or theoretical corrections have to be applied; alternatively, the method can be used for comparison purposes between optically similar powders.

403.4.6 *Disc photocentrifuges* are available from *Joyce Loebl* and *Brookhaven.* In this type of instrument, a suspension of particles at low concentration is floated on top of a clear liquid in a rotating disc. As the particles settle outward, they attenuate a white light beam at a fixed radius. As in the previous section, particle size is determined from the Stokes equation and the concentration undersize from the attenuation; the problems due to the breakdown in the laws of geometric optics apply.

403.4.7 *Cuvet photocentrifuges* are available from *Horiba, Seishin,* and *Shimadzu.* These operate using a homogeneous suspension as opposed to the line-start method described above. The centrifuge may be run at a constant speed or speed increasing with time (gradient mode) to reduce the analytical time. Correction is necessary to account for radial dilution, because the particles move in radially diverging paths as opposed to the parallel paths found in gravity sedimentation, as well as for the breakdown in the laws of geometric optics.

403.4.8 *Gravitational x-ray scanning sedimentation.* These devices are available from Micromeritics (*Sedigraph*), Quantachrome (*Microscan*), and *Brookhaven.* The particle size is determined using the Stokes equation, and the mass concentration undersize by the attenuation of a finely collimated x-ray beam. Analytical time is reduced by scanning either the beam towards the surface of the suspension or scanning the sedimentation cell past a fixed source and detector. The advantage of this instrument over photosedimentation is that the attenuation of the beam is directly proportional to the mass of powder in the beam — provided that the powder is homogeneous in composition. The disadvantage is that the method is only applicable to materials having an atomic number greater than about 13 because lighter elements are x-ray transparent.

403.4.9 The *Brookhaven X-ray Scanning Centrifuge* operates in either the gravitational or centrifugal mode. In the former mode, it operates in a similar manner to the instruments described above. In the centrifugal mode, the beam scans to the surface of a rotating disc in order to speed up the analysis, covering a 15:1 range in under 10 minutes. Blending of the gravitational and centrifugal data permits analyses from 100 μm to less than 0.05 μm.

403.4.10 *Low-angle laser light scattering* is provided by, *e.g.*, *Microtrac, Malvern, Shimadzu, Cilas, Coulter LS, Seishin, Horiba, Helos,* and *Fritsch.* When a particle is illuminated by a parallel beam of monochromatic light, a diffraction pattern is formed superimposed on its geometric image. An assembly of monosize spherical particles would give an enhanced image. A range of particle sizes can be divided into size intervals, each one of which will generate a diffraction pattern according to its average size — the intensity of which depends upon the number of particles in the size interval. Deconvolution of this pattern will generate the size distribution. This principle, combined with polarization and back scattering measurements using a secondary white light source, can cover the size range of 0.05 µm to over 1 mm.

403.5 *Particle Counting*

In particle counting, individual particles are sized and counted. This gives a number distribution from which a mass distribution can be calculated. As stated in Sect. 403.1, the technique suffers from a precision problem if a few large particles are improperly counted.

403.5.1 The *Electrical sensing zone method (Coulter Counter, Elzone)* is a technique whereby particles in a very dilute suspension are caused to pass through a small orifice on either side of which is immersed an electrode. The changes in electrical impedance as particles pass through the orifice generate voltage pulses whose amplitudes are proportional to the volume of the particles. The pulses pass to a pulse height analyzer and are classified into 16, 32, 64, 128, or 256 channels, depending on instrument and mode of operation, commonly in a 2:1 progression of discrimination, equivalent to a cube root of two progression in sizes, with a range of approximately 10,000 to 1. The resolution may be enhanced by a factor of ten in order to obtain precise information on very narrowly classified powders. A range of orifice sizes are available ranging from about 10 µm up to 2 mm; the 100-µm orifice covering the approximate volume diameter range of 43 µm to 2 µm. Multiple orifices can be used to cover wider size ranges. Because a mass balance can be carried out using this technique, the amount outside the measuring size range of the instrument may be estimated. Data also may be presented as number oversize per gram of powder on a log-log plot, which is useful for contamination studies. The method requires a conductive liquid and has a lower practical size limit of about 0.7 µm due to background electrical noise.

403.5.2 *Optical counters* fall into two categories: ***light blockage*** and ***light scattering***; further, the light can be from a white or laser light source, and the detector can be a photocell or photodiode. Various models are generally available, some for liquid-based systems and some for air-based ones. Liquid-based systems include contaminant monitors, aggressive liquid samplers, batch monitors, on-line samplers, and so on. Light-blockage systems are exemplified by the *Hiac/Royco,* which operates over a 45:1 size range from 0.25 μm and up for air, and from 0.5 μm and up for liquids, with an upper size of about 1 mm — these limits, however, are constantly being extended. The *Climet* CL200 series collect forward-scattered light with an elliptical mirror; the CL220 are liquid-borne analyzers; the CL221 is an on-line monitor; a light obscuration model is also available. Liquid borne instruments are also available from *Horiba, Insitec, Kane May, Kratel, Met One, Polytec, Procedyne,* and *Spectrex. Particle Measuring Systems* manufactures a range of laser based instruments. Further information on these and other instruments may be found in Ref. 803.18.

403.5.3 The *Brinkmann Analyzer* scans the sample with a focussed laser beam using a rotating-wedge prism. Particle size is interpreted by dwell time, *i.e.,* the shadow duration on a photodiode. The instrument operates in size ranges from 0.7 μm to 1,200 μm using a variety of measuring cells.

403.5.4 *Lasertec* uses a laser beam to scan over a small viewing area. Back-scattered light is picked up by a stereoscopic viewing system and particle size is deduced from dwell time. The suspension can be in a beaker or a tube (on-line), the infrared beam being focused about 1 mm inside the suspension. The size ranges covered vary from 0.5 μm to mm sizes.

403.5.5 The *Flowvision Analyzer* uses fiber optics to channel white light through a flowing liquid. As particles pass through the light beam, they generate images which are converted to video and then analyzed using a high speed digital computer. Particles are detected in the 2 μm to 1,000 μm size range with an optimum range for sizing of 25 μm to 600 μm. Suppliers are *Kevlin Microwave* and *Boston.*

403.6 *Surface Area Measurement*

Surface area is probably a powder's most important property because it controls the powder's rate of interaction with its surroundings. Although surface area does not give a size distribution, it is an integrated function of size distribution. Surface area is in itself of great interest in many applications such as grinding, pelletizing, catalysis and chemical reaction, and combustion.

403.6.1 The *BET* (Brunauer, Emmett, Teller) *nitrogen gas adsorption method* involves measuring the amount of gas that is adsorbed on a powder at liquid nitrogen temperature. Experimental data on the adsorbed volume is plotted against the pressure to generate the adsorption pressure isotherm and, using the BET theory, to calculate the amount of gas to give monolayer coverage. Based on theory, the area occupied by one molecule is determined and, from this, the surface area of the powder can be found. The powder is placed in a container which is exposed to a high vacuum at an elevated temperature to remove adsorbed gases. The container is then immersed in liquid nitrogen and exposed to gaseous nitrogen under known, increasing pressures. Simple gas theory permits the calculation of the amount of gas condensing on the powder as the relative pressure increases from 0.5 to 0.35 and, from this, BET theory gives the monolayer capacity. In an alternative procedure, a mixture of nitrogen and an inert gas (usually helium) flows over the powder and the amount adsorbed, at increasing relative pressures, is determined by gas chromatography. In the single-point technique, a less accurate assay is made at a single relative pressure (usually 0.2 or 0.3). Gases other than nitrogen have been used, particularly for low-surface-area powders, where greater accuracy is possible with krypton or argon due to their lower vapor pressures at liquid nitrogen temperature. The BET method is time consuming and involves relatively expensive equipment. Gas adsorption apparatus are available from *Carlo Erba, Chandler, Leeds and Northrup, Micromeritics, Netsch, Omicron, Quantachrome* and *Strohlein.*

403.6.2 In *air permeametry*, gas flow rate and pressure drop are measured through a packed bed of powder under ambient conditions at a known porosity and, from these measurements, a specific surface is obtained by semi-empirical methods. The equation governing flow through coarse powders is known as the *Carman-Kozeny equation* and must be modified by the use of a diffusion term for fine powders or coarse powders and gases at reduced pressures to generate the *Carman-Arnold equation.*

403.6.3 In the *Blaine Permeameter,* the inlet end of the bed is open to the atmosphere and air is drawn through the bed by the suction effect as an unbalanced column of oil comes to equilibrium. Because the pressure drop varies as the analysis proceeds, a modified permeametry equation is used. This is a constant volume permeameter, originally designed for cement, which operates at a fixed porosity thus limiting its applicability.

403.6.4 The *Griffin Permeameter* operates in a similar mode except that it is not a fixed porosity instrument, and the column of oil rebalances across the bed of powder in a closed system.

403.6.5 The *Fisher Sub-Sieve Sizer* is a constant pressure apparatus designed for routine analysis. One cubic centimeter of powder is packed to a known porosity, air is drawn through at a constant pressure drop, and the surface area determined from a self-calculating chart. By decreasing the porosity in steps, a reproducible measure of surface area is found.

403.6.6 Permeametry apparatus involve simple equipment which are easy to use. The *Fisher* and *Griffin* are more versatile because they can be used over a range of porosities. The size range (surface-volume mean diameters) is 0.02 μm to 200 μm — provided that the appropriate (Carman-Arnold) equation is used. The BET surface area is greater than the permeametry surface area, because the former surface includes cracks and fissures, whereas the latter only includes the envelope surface area of the particles. For this reason, the permeametry surface is commonly only about one third of the BET surface.

500.0 **Test Procedure**

501.0 *Introduction*

Once the objectives of the test have been determined, as discussed in Sect. 300.0, then the specific test procedure can be decided upon. A test procedure includes the following steps.

502.0 *Method of Feeding*

502.1 *Fluctuation in feed conditions* can be an important source of error in measuring classifier performance. The reason is that the particles reside in the classifier for a finite time so that product sampled at one instant of time may represent a shift in feed particle size distribution from a previous instant.

502.2 It is important that the *feed rate* to a classifier be maintained constant. Feeders that limit feed rate variation to ±5% are recommended. Ref. 803.9, Sect. 7, p. 4, describes various feeders available for dry solids.

502.3 *Slurries* may be constituted from dry powder streams mixed with liquid streams. In this way, the flow rate of each stream can be measured separately. In feeding a slurry from a mixing tank, one should check the uniformity of concentration and particle size delivered over time.

503.0 *Duration of Test Period*

503.1. The outputs of the classifier will vary with *fluctuations* in the feed rate and feed composition caused by variations in an upstream process or in the feeding mechanism. The extent of such fluctuations can be gauged by the *control chart method* of Shewhart, as described in Ref. 803.16. Use of this method requires measurement of the important feed variables such as flow rate and size distribution over several periods of time to determine whether there are fluctuations exceeding expected values and a trend of deviations indicating a lack of control. Steady operation should be maintained for a long enough period before sampling begins to assure that a steady state has been reached. If the feed conditions do not vary significantly, then steady operation of the classifier itself will require at least five times the residence time of the particles in the system.

503.2 In addition, each sampling should be carried out over a sufficient time to assure that fluctuations in solid and fluid flow rates, etc., are averaged out, based on the times for steady operation measured as described above.

504.0 *Method of Measuring Flow Rates*

504.1 It generally is necessary to measure both solid and fluid flow rates or to measure that of the suspension and its solids concentration.

504.2 Quantities of *dry feed material* can be determined by measuring the weight passing over a *belt*, the reduction of weight in a *feed hopper*, or by the increase of weight in a *product hopper*. Samples can be collected over a period of time and weighed, but this may be impractical if the flow rate is large.

504.3 If the stream is a *slurry*, its flow rate can be measured by a *flow meter*, but the potential of plugging must be considered (Ref. 803.9, Sect. 5, p. 17). The proportion of solids and liquid can then be determined by taking a sample and measuring its *density*, or by evaporating it to dryness and weighing the solids remaining. There are also various *nuclear gauges* used to measure the mass flow of slurries or of dry solids as they pass by at a known velocity.

504.4 If the particles are suspended in a *gas stream*, similar methods can be used to measure particle and fluid flow rates. In this case, *particle concentration* can be measured by methods discussed in Sect. 402.6. Gas flow rates can be measured by *pitot tubes, orifice meters*, or other methods described in Ref. 803.9, Sect. 5, pp. 7–17.

504.5 In setting up a test, provision should be made to measure flow rates of feed and both fraction streams. A *reconstituted feed stream* can then be calculated based on the measured fraction streams. Comparison with the measured feed stream will allow a check of error by material balance. In cases where one stream is inaccessible, its size distribution can be deduced from the other two — this practice is not recommended, however, because it gives no indication of error.

505.0 *Method of Sampling and Subdividing Samples for Analysis*

The purpose of sampling procedures is to acquire a *representative sample* for size analysis or composition analysis. One of the greatest sources of error in measuring the performance of particulate systems is that of obtaining representative samples. Whenever particles of various sizes, shapes, and densities are handled, vibrated, or conveyed, *segregation* by size, shape, or density can occur. Acceptable sampling techniques are designed to minimize the effects of segregation. Preferred methods are described in Sect. 402.4. These methods are based on devices that will cut portions from the whole stream at successive time intervals, thus avoiding segregation by position, and averaging variations over time.

506.0 *Dispersion of Particles*

506.1 It is important that particles be well *dispersed*, both for accurate particle size analysis and for proper operation of the classifier. (See Sect. 703.6.) The smaller the particles, the more important and difficult achieving good dispersion will be.

506.2 *Methods of dispersing particles in liquids* are well known — for example, see Ref. 803.12. The general procedure is to agitate the suspension to break up agglomerates, and to adjust the zeta potential so that electric forces cause particles to repel each other and remain dispersed. This can be done by adjusting pH or adding dispersing agents.

506.3 *Methods of dispersing particles in an aerosol* are not as well known. Simply feeding the particles directly into an air stream may suffice if the feed is free flowing. If not, severe agitation may be required to break up agglomerates. This could, for example, be achieved in a continuous air swept hammer mill, if the resulting particle size reduction is appropriate. *Reagglomeration* can be prevented by allowing insufficient time for agglomeration, and by keeping the particle concentration low enough so that the rate of interparticle collision is small. Ref. 803.14 indicates that the mean free path of particles in an aerosol is inversely proportional to the number concentration of particles, as it is in the kinetic gas theory.

For aerosols, electric forces are not usually used to aid dispersion in the way zeta potential is used for liquid dispersions, although electric forces do affect agglomeration. *Additives* to powders or *surface chemical treatments* can improve *dispersibility* by reducing adhesive forces between particles, but they do not cause dispersion by themselves, nor do they stabilize the dispersion.

506.4 To evaluate a classifier for a particular application, it is only necessary that the *degree of dispersion* of the feed stream represent the conditions of feed in the application. See also Sect. 702.

507.0 *Method of Particle Size Analysis*

No one method can be recommended in general. Principles to be applied in selecting the method are discussed in Sect. 401.2. Methods of particle size analysis are described in Sect. 403.0. Many of the leading methods are described, with additional references given.

508.0 *Statistical Control of Precision and Accuracy*

508.1 The *control chart method* mentioned in 503.1 can also be applied to variables other than time sequences, as explained in Ref. 803.16, to determine whether there are uncontrolled *systematic errors* in experimental methods, especially the method for measuring particle size distribution. If the precision of the laboratory performing the particle size analysis is not already known, it is recommended that a series of measurements be made on representative samples of the feed to determine whether there is any uncontrolled variation or trend in the analysis results. This can be done, for example, by submitting occasional replicate samples throughout the test series.

508.2 The *accuracy of measurements*, particularly of particle size analyses, should be determined beforehand so a decision can be made whether the accuracy is sufficient for the intended purpose of the work. Accuracy of particle size analysis methods can be determined by calibration with *standard reference materials* (SRMs), as described in Ref. 803.17. Examples of such materials are the National Institute of Standards and Technology's SRMs 1690, 1691, etc., and the British BCRs 165, 166, 167, or the BCR quartz series 66, etc. In most cases, the manufacturer of the particle size analysis equipment can supply an estimate of the achievable accuracy and the method by which it was calibrated.

509.0 *Test Data to be Recorded*

The recorded classifier test data should be complete enough to permit test duplication by others, and to assure accurate scaleup and related cost estimates. A summary of items to be considered is given in Sect. 303.0. Data sheets should be prepared beforehand to assure that all necessary measurements are recorded.

510.0 *Observations on Equipment Operability*

510.1 The *operability* of the classifier may have a dominant effect on operating *cost.* The following examples of special items should be observed in any test program:

- evidence of classifier corrosion or wear;
- unusual motor load fluctuation and its cause;
- a tendency for plugging in the feed inlet or fine fraction outlet;
- the time required for complete cleanout when a different product is to be produced;
- the convenience of changing particle cut size; and
- product contamination.

510.2 Some *low-melting* or *waxy materials* tend to smear on internal surfaces and, in time, may plug internal classifier parts. The condition may not be apparent in a short test, but careful post-test inspection of internal parts should reveal if a long term test run is warranted.

600.0 **Computation of Results**

601.0 *Intrinsic Classifier Performance*

601.1 *Parameters to be Considered*

Basically, three parameters should be focused on when evaluating the performance of a given classifier: *cut size, sharpness of cut* (because these affect the product particle size distribution), and *processing capacity*. Cut size is essential because it represents the particle size at which classification actually takes place, while sharpness represents how effectively it is done. Capacity is important because it directly relates to finished product cost. With most conventional classifier designs, sharp separations at smaller cut sizes are achieved only at the expense of reduced capacity. The significance of these parameters will be discussed more fully in Sect. 700.0.

601.2 *Size Selectivity*

Size selectivity is the most complete method of expressing classifier performance under a given set of operating conditions. Cut size and sharpness can be calculated from size selectivity data. Size selectivity, expressed as a fraction, (see Sect. 205.6) is defined by:

s_D = quantity of size D entering coarse fraction/quantity of size D in feed

where s_D is classifier selectivity and D is particle size. The equivalent mathematical expression is, on a mass basis:

$$s_D = w_c\, d\phi_c / w_o\, d\phi_o = w_c\, d\phi_c / (w_c\, d\phi_c + w_f\, d\phi_f) \qquad (6.1)$$

where ϕ_c is the cumulative percent by mass of coarse fraction less than particle size D, ϕ_f is the cumulative percent by mass of fine fraction less than particle size D, ϕ_o is the cumulative percent by mass of feed less than particle size D, w_c is the coarse fraction mass flow rate, w_f is the fine fraction mass flow rate, and w_o is the feed mass flow rate.

601.3 *Mass Balance Considerations*

It should be noted in Equation (6.1) that the masses of feed and both fractions balance. It is generally preferable to calculate the feed size distribution from measured size distributions of both coarse and fine fractions than to measure the feed and one of the fractions and calculate the other fraction. If all three are measured, it is best to use the feed size analysis only for assessment of measurement error by material balance. These recommendations follow from (1) the fact that size analyses are subject to less error the more uniform the fraction; and (2) the possibility of loss or grinding in the classification process. The cumulative feed size distribution, ϕ_o, is calculated from cumulative coarse and fine fraction size distributions, ϕ_c and ϕ_f, as follows:

$$\phi_o = (w_c/w_o)\,\phi_c + (w_f/w_o)\,\phi_f \qquad (6.2)$$

601.4 *Size Selectivity Calculation*

601.4.1 For purposes of calculating size selectivity from cumulative particle size distribution data, Equation (6.1) can be expressed in incremental form as follows:

$$s_{D_i} = w_c\,\Delta\phi_{ci} \,/\, (w_c\,\Delta\phi_{ci} + w_f\,\Delta\phi_{fi}) \qquad (6.3)$$

where, as defined in Figure 6.1, $\Delta\phi_{ci}$ and $\Delta\phi_{fi}$ are, respectively, the cumulative size distribution intervals of coarse and fine fractions associated with size interval ΔD_i. An interval-representative size, D_i, is arbitrarily taken as the midpoint of ΔD_i.

601.4.2 The *precision* attainable in the use of Equation (6.3) is dependent on how narrow ΔD_i is chosen. It is not necessary to select a constant value of ΔD_i in calculating size selectivity. A good rule to follow is that ΔD_i should be less than 20% of D_i While ΔD_i should be taken as small as possible to give a mathematically correct measure of s_{D_i}, it is rarely necessary to have ΔD_i as small as 10% of D_i.

One way to achieve an appropriate division is to choose sizes that result in 10 to 15 fractions, with narrower fractions at the ends, especially the fine end of the distributions. For example, one could

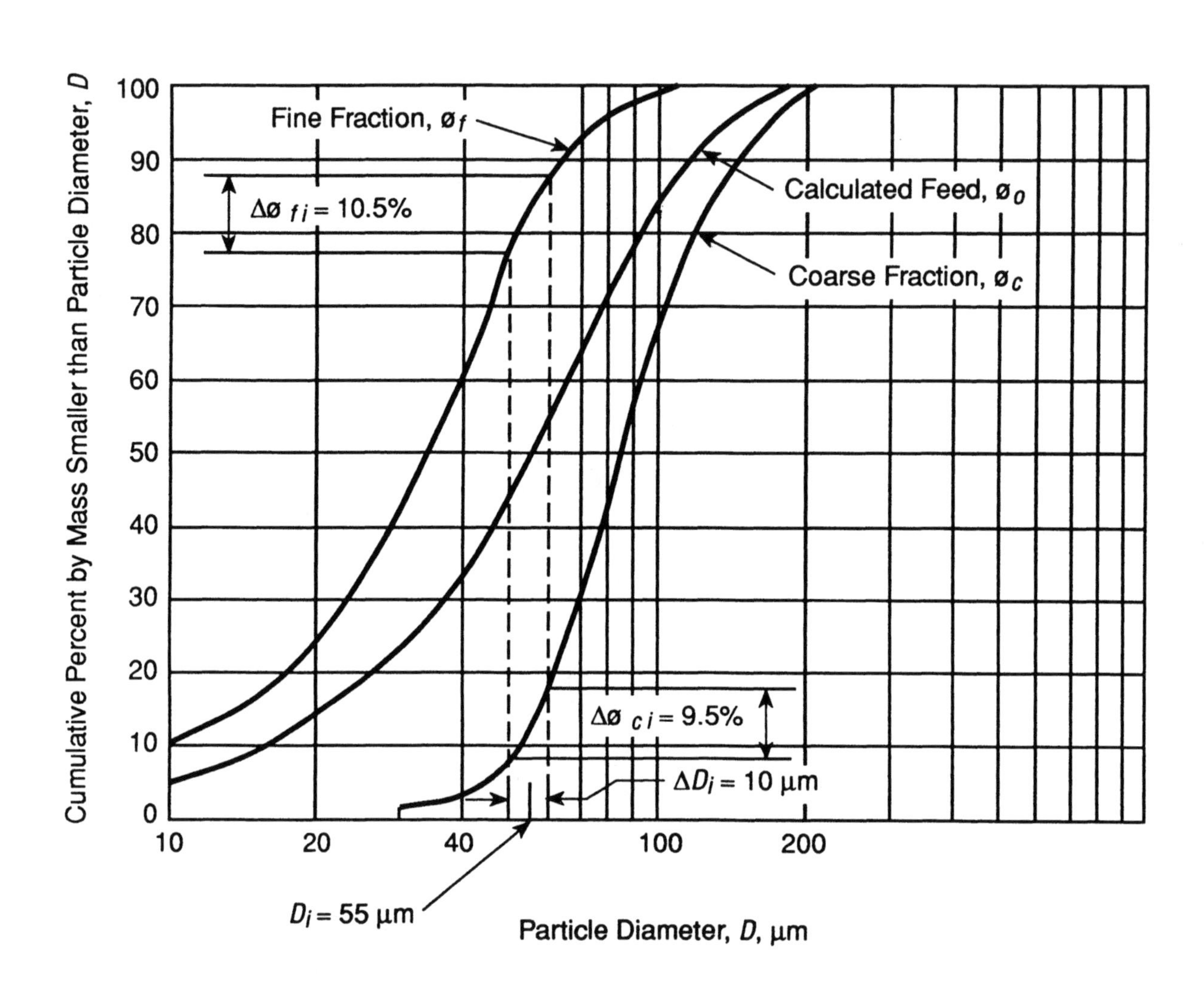

Figure 6.1. Example calculation of size selectivity from cumulative particle size distributions.

target the following weight percentages counting from the fine end: 1, 2, 2, 5, 10, 10, 10, 10, 10, 10, 10, 10, 5, 5. To find the sizes corresponding to these divisions, one can make a preliminary experiment — analyzing the products into 5 fractions. From this information, one can then choose sizes to get 10 to 15 narrower fractions for a more definitive experiment.

601.4.3 A *sample calculation* of size selectivity from cumulative particle size distribution data using Equation (6.3) is given in Sect. 802.0. This is based on the data given in Figure 6.1 for a centrifugal air classifier operating such that w_c = 0.46 ton/hr (0.12 kg/sec) and w_f = 0.54 ton/hr (0.14 kg/sec). The size selectivity curve thus obtained is shown in Figure 6.2. This method lends itself to rapid, routine calculation using computer techniques. Graphical methods can also be used for direct calculation of size selectivity from cumulative particle size distribution data (Ref. 803.4).

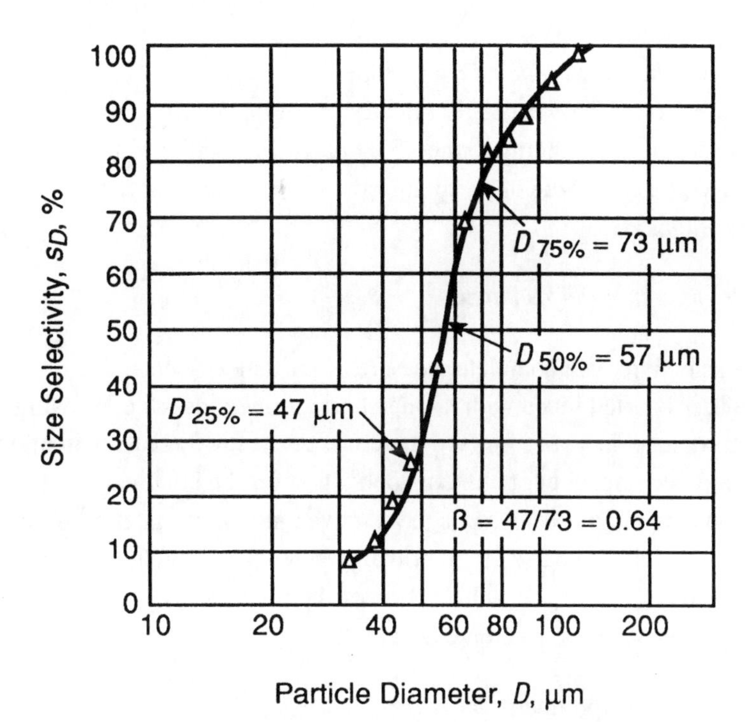

Figure 6.2. Size selectivity for example calculation.

601.5 *Performance Parameters Derived From Size Selectivity*

601.5.1 Figure 6.2 exemplifies a calculated size selectivity curve or function. If the separation were ideal, this would consist of a straight vertical line at the *equiprobable cut size* $D_{50\%}$. This means that all particles in the feed larger than $D_{50\%}$ enter the coarse fraction, and all particles smaller than $D_{50\%}$ enter the fine fraction. For an actual classification, the curve varies from vertical by some amount, indicating the extent to which particles of each size are misplaced. From the definition of equiprobable cut size, $D_{50\%}$ corresponds to the 50% value on the size selectivity curve.

601.5.2 The shape of the selectivity curve at $D_{50\%}$ indicates the spread of misplaced material or *sharpness of cut.* There are many ways in which sharpness can be expressed. One index related to the selectivity curve shape that has been widely used is the ratio:

$$\beta = D_{25\%}/D_{75\%} \tag{6.4}$$

where β is the *sharpness index*, $D_{75\%}$ is the particle size corresponding to the 75% classifier selectivity value, and $D_{25\%}$ is the particle size corresponding to the 25% value. For perfect classification, β has a value of unity; the smaller β, the poorer the sharpness of classification.

602.0 *Overall Classification Performance*

602.1 While size selectivity is a complete measure of particle size classifier performance, the user is often faced with taking shortcut methods for expressing classification performance on a specific feed material. A practical measurement of overall classification performance for a given application can be obtained by calculating recovery and yield. *Recovery* is the relative amount of material in the feed of the desired size (either coarser or finer than a given size *D*) that is recovered in the product. Recovery, expressed as a fraction of the feed, can be calculated from cumulative particle size distribution data as follows:

When the *fine fraction* is the product:

$$R_{D_f} = w_f \, \phi_f / w_o \, \phi_o \tag{6.5}$$

When the *coarse fraction* is the product:

$$R_{D_c} = w_c\,(1 - \phi_c)\;/\;w_o\,(1 - \phi_o) \qquad (6.6)$$

Yield is a measure of product obtained irrespective of quality, and is calculated as a fraction of feed as follows:

Fine yield:

$$Y_f = w_f\;/\;w_o = 1 - Y_c \qquad (6.7)$$

Coarse yield

$$Y_c = w_c\;/\;w_o = 1 - Y_f \qquad (6.8)$$

For cases in which w_c and w_f cannot be measured, yield can be estimated from size distributions if mass balance is assumed:

$$Y_c = (\phi_f - \phi_o)\;/\;(\phi_f - \phi_c) \qquad (6.9)$$

and

$$Y_f = (\phi_o - \phi_c)\;/\;(\phi_f - \phi_c) \qquad (6.10)$$

602.2 *The particle size at which recovery is evaluated* is determined by the application. For example, if the desired product in Figure 6.1 is the fine fraction below a particle size of 70 μm, the recovery based on the calculated feed size distribution would be:

$$R_{70\mu m} = (0.54 \text{ ton/hr})(93\%)/(1.00 \text{ ton/hr})\ (64.7\%) = 0.78$$

This means that only 78% of the desired feed material shows up in the fine fraction. The fines yield in this example is simply 0.54 (or 54%).

602.3 Although simple to calculate and useful if properly interpreted, both yield and recovery are related to a given application, and are not unique measures of classifier performance.

603.0 *Other Measures of Classification Performance*

603.1 *Classifier "Efficiency"*

While the classifier selectivity curve gives an assessment of classifier performance, the user is commonly concerned with classification performance on a specific feed material. Consequently, various *single number measures* of classification have come into use. Often referred to as classifier "efficiencies," these have been extensively reviewed in Refs. 803.10 and 803.11. (See the footnote on p. 17.) In most cases, such measures tend to be geared to a particular application. Moreover, it can be shown that all single number efficiencies are dependent on the following factors:

- the classifier size selectivity;
- the feed material particle size distribution; and
- the particle size at which performance is evaluated.

It is for this reason that the *preferred procedure* for reporting performance is in terms of the classifier selectivity function, rather than in terms of a single number parameter alone. If the user must use a shortcut method for measuring overall classification result, the more generally applicable recovery calculation given in Sect. 602.0 is recommended. While recovery is dependent on size selectivity and feed powder particle size distribution, it does have an obvious physical significance and gives a direct measure of classification for a given material in the general case. Thus, when properly interpreted, recovery gives information that is not directly apparent from the selectivity curve.

603.2 *Specific Surface Method*

603.2.1 The specific surface average size†, D_s, of either the coarse or fine fractions can be used as a measure of classification effectiveness. For a given fractional separation of a material, there is a given specific surface size for each fraction that corresponds to a perfect classification. As classification becomes poorer, the actual specific surface size of the coarse and fine fractions approach each other and that of the feed. The specific surface size of the feed

† This size, D_s, is often called the Sauter diameter and designated D_{32} (Ref. 803.8)

material and the coarse and fine fractions are related by a material balance and, for the case where the specific gravity of all fractions is the same, this is given by:

$$D_{so} = 1 / [(Y_f / D_{sf}) + (Y_c / D_{sc})] \qquad (6.11)$$

where D_{so} is the particle diameter having the same specific surface as the feed as a whole, D_{sc} is the particle diameter having the same specific surface as the coarse fraction as a whole, and D_{sf} is the particle diameter having the same specific surface as the fine fraction as a whole.

603.2.2 The ratio D_{sc} / D_{sf} becomes an exceptionally sensitive measure of effectiveness of classification for a given feed material. This ratio, however, will be a function of the size distribution of the feed material and will not be an intrinsic property of the classifier. The specific surface average size, D_s, is obtainable from the measured specific surface by:

$$D_s = 6 / \rho S \qquad (6.12)$$

where ρ is the true particle density, and S is the specific surface of the powder on a mass basis (m^2/kg).

700.0 **Interpretation of Results**

701.0 *Introduction*

Classifier performance is determined partly by *classifier design*, and partly by *feed* and *operating conditions*. For a given classifier and feed material, sharpness of separation is determined by the particle feed rate — whereas the nominal size of the product is determined by the cut size for which the operating conditions of the classifier are set. If the product can tolerate more of the undesired fraction and still be within desired quality specifications, *yield* can be increased by changing operating conditions so as to give a different cut size (larger if the fine fraction is the desired product; smaller if the coarse fraction represents the desired product). In a dedusting operation, it is expedient to operate at as large a cut size as possible without removing a significant amount of material from the coarse fraction.

702.0 *Operating Conditions*

702.1 The *maximum sharpness*, which is limited by the intrinsic characteristics of the classifier, will be achieved at very low particle feed rates. As feed rate is increased without changing other operating conditions (such as fluid flow rates, rotor speeds, or vane settings), production rate will increase; sharpness of separation, however, will become poorer. This is partly the result of poorer dispersion of the particles either because of poorer initial dispersion; the greater probability of interaction or agglomeration between particles at the higher particle concentrations; hindered motion of particles at higher concentrations; or increased nonideal flow patterns, such as secondary flows or eddies. See also Sects. 506. and 703.6.

702.2 With the *particle loadings* employed in air classification (under ⅔ lb/ft^3 or approximately 10 kg/m^3), *drag forces* on individual particles are essentially the same as those that would be exerted on the particles in infinite space (so-called "free settling"). For very high volumetric solid concentrations (over ⅔ lb/ft^3 or approximately 10 kg/m^3) such as might be encountered in some liquid suspensions, drag forces on individual particles are larger than they would be for particles in infinite space due to the proximity of other particles (so-called "hindered settling"). Thus, at these higher concentrations, *nonideal behavior* may be expected, and performance will differ from intrinsic classifier performance. This will affect recovery as well as selectivity. It will be possible to compensate for a change in *recovery* with increased loading by changing operating conditions to change the cut size, but this will not compensate for a change in *sharpness of separation*.

703.0 *Size Selectivity Curves*

703.1 Figure 7.1 depicts the various types of size selectivity curves that are possible or may be obtained. The following paragraphs will discuss the source or significance of each type of behavior. In Figure 7.1, the same cut size, $D_{50\%}$, has arbitrarily been chosen for all curves. It is also assumed in Figure 7.1 that the size selectivity curves are derived from size analyses that have correctly measured the discrete particle diameters in the pertinent fractions — which assumes that particles not dispersed in the classifier have been dispersed in the size analysis. If particle flocculation complicates the size analyses, a wide variety of erratic, meaningless results could be obtained.

703.2 Curve a-a' is representative of a *perfect separation.* Curve b-b' is representative of the *intrinsic capability* of a given classifier and is the best that can be achieved with that classifier at low particle loadings.

703.3 Curves b-c', c-b', c-c', d-b', and d-c' are representative of the apparent size selectivity results that actually may be obtained as the result of additional complications introduced by the mode of classifier operation or particle interaction. *Deviations* from a curve like b-b', which passes through 0% at the finest particle size end and 100% at the coarse end, may be due to one of the following:

- mechanical loss of material during the classification process;
- grinding of material during classification;
- recycle streams within the classifier;
- bypass resulting from splitting main fluid carrier stream, *i.e.*, some of the feed passes directly into the product; or
- failure to disperse or flocculation of particles during classification.

703.4 If the size analysis of the feed calculated from the size analyses of the coarse and fine fractions agreed with the actual size analysis of the feed, the same size selectivity curve will be obtained regardless of which combination of fractions is used in the calculation of the size selectivity curve. If the calculated feed analysis does not agree with the measured feed, a different size selectivity curve will be obtained depending on which combination of size analyses is used to calculate the size selectivity curve.

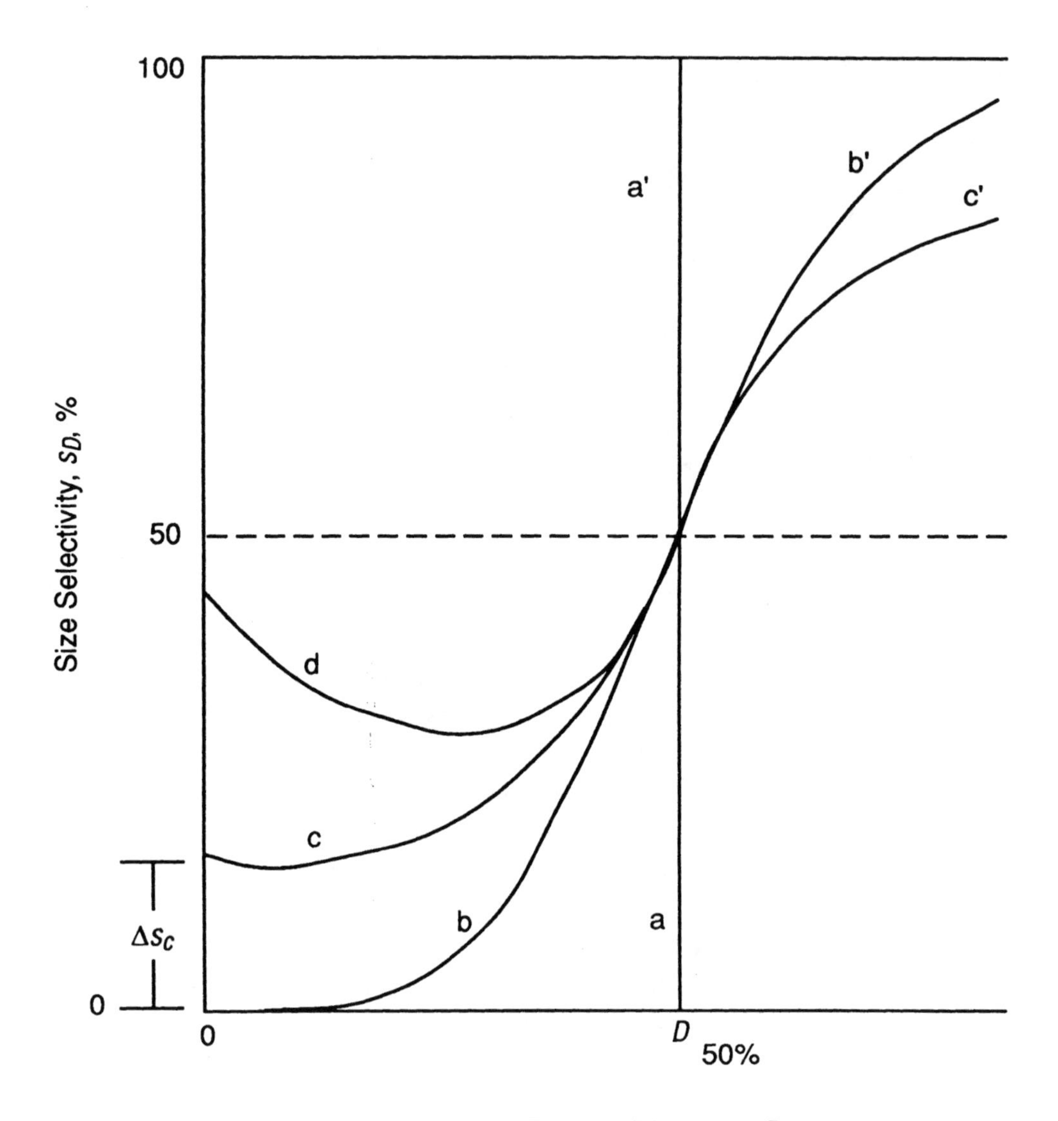

Figure 7.1. Types of apparent size selectivity curves.

703.5 Curve c' could reflect the effects of *grinding*. This could be confirmed if the feed analysis calculated from the analyses of the fine and coarse fraction is finer than that measured for the feed.

703.6 The fishhook curve d is indicative of a failure to adequately *disperse* the feed material. The agglomerated fines act like coarse particles in the classifier. This condition will be detected only if the agglomerates are redispersed in the particle size analyzer.

703.7 The leveling off shown by curve c may be due either to contamination of the fine fraction by *recycle* streams or failure to *disperse* the feed. In general, material balances on both an overall mass and on a particle size basis should reveal which phenomena (material loss, recycling, grinding, or maldispersion) are responsible for deviations from a curve of the shape of b—b' in any specific case.

703.8 In the case of curve c-c', a *corrected curve* could be calculated to yield one of the form of b-b' from the following relationship (Ref. 803.7):

$$s_{cDi} = (s_{Di} - \Delta s_c) / (1 - \Delta s_c - \Delta s_{c'}) \quad (7.1)$$

where s_{Di} is the measured size selectivity value, s_{cDi} is the corrected value, and Δs_c and $\Delta s_{c'}$ are the size selectivity increments at the fine and coarse ends, respectively, as illustrated in Figure 7.1. This relationship simply imposes the proper boundary values on the data. While this corrected curve is of interest for purposes of modeling classifier performance to account for effects such as bypassing, the true correction could be considerably more involved depending on the reasons for the deviations in the measured values.

800.0 **Appendices**

801.0 *Notation*

D	= particle size
D_i	= midpoint of size interval ΔD_i
ΔD_i	= particle size interval
D_s, D_{32}	= Sauter diameter, *i. e.*, particle diameter having the same specific surface as the sample as a whole
D_{sc}	= particle diameter of coarse fraction having the same specific surface as the coarse fraction as a whole
D_{sf}	= particle diameter of fine fraction having the same specific surface as the fine fraction as a whole
D_{so}	= particle diameter of feed having the same specific surface as the feed as a whole
$D_{25\%}$	= particle size corresponding to the 25% size selectivity value
$D_{50\%}$	= equiprobable cut size, *i. e.*, the particle size corresponding to the 50% size selectivity value
$D_{75\%}$	= particle size corresponding to the 75% size selectivity value
R_{Dc}	= coarse fraction recovery, *i. e.*, amount of desired coarse material expressed as a fraction of the feed
R_{Df}	= fine fraction recovery, *i. e.*, amount of desired fine material expressed as a fraction of the feed
s_D, s_{Di}	= size selectivity
s_{cDi}	= "corrected" classifier size selectivity
S	= specific surface of a powder on a mass basis
w_c	= coarse fraction mass flow rate
w_f	= fine fraction mass flow rate
w_o	= feed mass flow rate
Y_c	= coarse fraction yield expressed as a fraction of the feed
Y_f	= fine fraction yield expressed as a fraction of the feed
β	= sharpness index, *i. e.*, $D_{25\%} / D_{75\%}$
ρ	= true particle density

ϕ_c = cumulative percent by mass of coarse fraction less than particle size D

ϕ_f = cumulative percent by mass of fine fraction less than particle size D

ϕ_o = cumulative percent by mass of feed less than particle size D

$\Delta \phi_{ci}$ = cumulative size distribution interval of coarse fraction associated with size interval ΔD_i

$\Delta \phi_{fi}$ = cumulative size distribution interval of fine fraction associated with size interval ΔD_i

802.0 *Sample Calculations*

802.1 An example calculation using data from an actual test is presented in Table 8.1. Mass flow rates and cumulative size distributions for coarse and fine fractions are as follows:

- $w_f = 0.54$ ton/hr (0.14 kg/sec);
- $w_c = 0.46$ ton/hr (0.12 kg/sec);
- $w_o = w_f + w_c = 1.0$ ton/hr (0.26 kg/sec); and
- ϕ_c and ϕ_f as functions of particle size from Figure 6.1.

802.2 It is desired to calculate the cumulative size distribution of the feed, ϕ_o, and the size selectivity curve for the process, s_{Di}, as functions of particle size. The feed size distribution is calculated using Equation (6.2) and graphed in Figure 6.1 (see Sect. 601.3):

$$\phi_o = (w_c / w_o)\, \phi_c + (w_f / w_o)\, \phi_f \qquad (6.2)$$

Size selectivity is calculated using Equation (6.3) and graphed in Figure 6.2 (see Sect. 601.4):

$$s_{D_i} = w_c\, \Delta\phi_{ci} \,/\, (w_c\, \Delta\phi_{ci} + w_f\, \Delta\phi_{fi}) \qquad (6.3)$$

803.3 From Figure 6.2, the equiprobable cut size, $D_{50\%}$, for this process is 57 µm, with a sharpness index calculated via Equation (6.4) as follows (see Sect. 601.5):

$$\beta = D_{25\%}/D_{75\%} = 47/73 = 0.64$$

W_c = 0.46 Ton/Hr (0.12 kg/sec) W_f = 0.54 Ton/Hr (0.14 kg/sec) $W_o = W_c + W_f$ = 1.00 Ton/Hr (0.26 kg/sec)

D µm	$D/5$ µm	Int. No i	ΔD_i µm	Midpoint D_i µm	$\varnothing_c$ %<D	$\varnothing_f$ %<D	Calc $\varnothing_o$ %<D	$\Delta\varnothing_{ci}$ %	$\Delta\varnothing_{fi}$ %	$W_c\Delta\varnothing_{ci}$ $(\frac{ton}{hr})(\%)$	$W_f\Delta\varnothing_{fi}$ $(\frac{ton}{hr})(\%)$	$W_c\Delta\varnothing_{ci} + W_f\Delta\varnothing_{fi}$ $(\frac{ton}{hr})(\%)$	S_{Di} %
30	6				0.8	44.0	24.0						
		1	4	32				0.7	6.6	0.32	3.56	3.88	8.2
34	6.8				1.5	50.6	28.0						
		2	6	37				1.5	10.9	0.69	5.89	6.58	10.5
40	8.0				3.0	61.5	34.6						
		3	4	42				1.7	6.5	0.78	3.51	4.29	18.2
44	8.8				4.7	68.0	38.9						
		4	6	47				3.8	9.5	1.75	5.13	6.88	25.4
50	10				8.5	77.5	45.8						
		5	10	55				9.5	10.5	4.37	5.67	10.04	43.5
60	12				18.0	88.0	55.8						
		6	10	65				13.2	5.2	6.07	2.81	8.88	68.4
70	14				31.2	93.2	64.7						
		7	10	75				12.6	2.4	5.80	1.30	7.10	81.7
80	16				43.8	95.6	71.8						
		8	10	85				11.7	2.0	5.38	1.08	6.46	83.3
90	18				55.5	97.6	78.2						
		9	10	95				11.0	1.4	5.06	0.76	5.82	86.9
100	20				66.5	99.0	84.1						
		10	20	110				13.5	0.9	6.21	0.49	6.70	92.7
120	24				80.0	99.9	90.7						
		11	20	130				8.0	0.1	3.68	0.05	3.73	98.7
140					88.0	100.0	94.5						

Table 8.1. Sample calculation for centrifugal air classifier.

803.0 *References*

803.1 "Efficiency Testing of Air-Cleaning Systems Containing Devices for Removal of Particles," American National Standard N101, American National Standards Institute, New York (1972).

803.2 "Methods for the Determination of Particle Size of Powders, Part 1, Subdivision of Gross Samples down to 2 ml." British Standard 3406, British Standards Institution, London (1961).

803.3 "Engineered Automatic Sampling Systems," Bulletin S-069, The Galigher Co., Salt Lake City, UT.

803.4 **Gibson, K.,** "Particle Classification Efficiency Calculations by Geometry," *Powder Technology,* **18,** pp. 165–170 (1977).

803.5 **Kaye, B. H.,** "Efficient Sample Reduction by Means of a Riffler Sampler, from Analysis of Calcareous Materials," Publication No. 18, pp. 159–167, Society of the Chemical Industry, London (1965).

803.6 **Lapple, C. E.,** "Particle Size Analysis and Analyzers," *Chem. Eng.,* **75** (11), pp. 149–156 (May 20, 1968); also appears in "Encyclopedia of Science and Technology," 3rd ed., **9,** pp. 662–669, McGraw Hill, New York (1971).

803.7. **Luckie, P. T., and L. G. Austin,** "Technique for Derivation of Selectivity Functions from Experimental Data," in proceedings of "10th International Mineral Processing Congress (1973)," Jones, M. J., ed., pp. 773–790), I. M. M., London (1975).

803.8 **Mugele, R., and H. Evans,** *Ind. Eng. Chem.,* **43,** p. 1,317 (1951).

803.9 "Chemical Engineers' Handbook," 6th ed., Perry, R. H., and D. Green, eds., McGraw Hill, New York (1983).

803.10 **Richards, J. C.,** "The Efficiency of Classifiers," *Monthly Bulletin,* **30,** pp. 113–139, British Coal Utilization Research Association (1966).

803.11 **Schultz, N.,** "Separation Efficiency," *American Institute of Mining Engineers Transactions,* **247,** pp. 81–87 (March 1970).

803.12 **Wnek, W.,** "Electrokinetic and Chemical Aspects of Water Filtration," *Filtration and Separation,* **11** (3), pp. 237–242 (1974).

803.13 **Box, G. E. P., W. G. Hunter, and J. S. Hunter,** "Statistics for Experimenters," Wiley, New York, (1978).

803.14. **Rumpf, H.,** "Loading Theory of Impact Size Reduction," *Chemie-Ing.-Techn.,* **31**, pp. 323–327 (1959). [National Translation Center translation 61-12395, Library of Congress, Washington.]

803.15. "Classifier Investigations," *Verein Deutscher Zementwerke,* Notice MT28 (Dec. 1965).

803.16. **Wernimont, G.,** "Statistical Control of Measurement Processes," in "Validation of the Measurement Process," DeVoe, J.R., ed., *ACS Symposium Series,* **63**, pp. 1–29, Washington (1978).

803.17 **Uriano, G.A., and J. P. Cali,** "Role of Reference Materials and Reference Methods in the Measurement Process," *ibid.*, pp. 140–161.

803.18 **Allen, T.,** "Particle Size Measurement," 4th ed., Chapman and Hall, London (1990).

803.19 "British Standard Method for Determination of Particle Size Distribution by Gravitational Liquid Sedimentation Methods," British Standard 3406, part 2, British Standards Institution, London (1984).

803.20 "British Standard Method for Determination of Particle Size Distribution by Centrifugal Liquid Sedimentation Methods for Powders and Suspensions," *ibid.*, part 6 (1985).

804.0 **Index**

Printed and bound by CPI Group (UK) Ltd, Croydon, CR0 4YY

17/04/2025

14658904-0001